AF270643

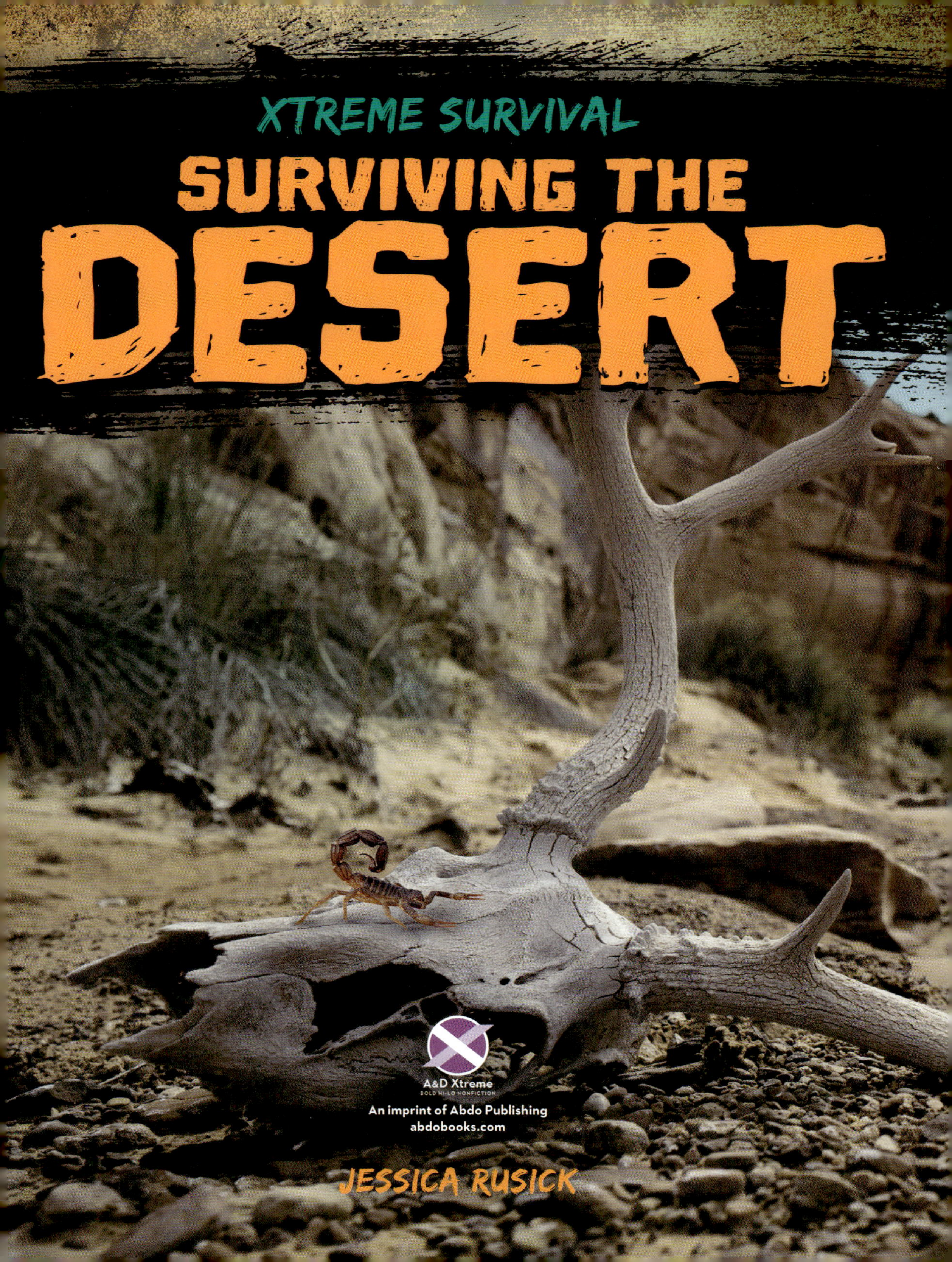

XTREME SURVIVAL
SURVIVING THE
DESERT
A&D Xtreme
BOLD HI-LO NONFICTION
An imprint of Abdo Publishing
abdobooks.com
JESSICA RUSICK

TAKE IT TO THE XTREME!

GET READY FOR AN EXTREME ADVENTURE!
THE PAGES OF THIS BOOK WILL TAKE YOU INTO
THE THRILLING WORLD OF OUTDOOR SURVIVAL.
WHEN YOU HAVE FINISHED READING THIS BOOK, TAKE THE
XTREME CHALLENGE ON PAGE 45 ABOUT WHAT YOU'VE LEARNED!

ABDOBOOKS.COM

Published by Abdo Publishing, a division of ABDO, PO Box 398166, Minneapolis, Minnesota 55439.
Copyright ©2024 by Abdo Consulting Group, Inc. International copyrights reserved in all countries.
No part of this book may be reproduced in any form without written permission from the publisher.
A&D Xtreme™ is a trademark and logo of Abdo Publishing.
Printed in the United States of America, North Mankato, MN.
102023
012024

Design: Tamara JM Peterson, Mighty Media, Inc.
Editor: Katherine Chu
Cover Photographs: EcoPrint/Shutterstock Images (scorpion); jrituc-ci/Shutterstock Images (desert canyon)
Interior Photographs: Aleksandr Melnikov/Shutterstock Images, pp. 18–19; Alexxxey/Shutterstock Images, pp. 14–15; anne-tipodees/Shutterstock Images, pp. 24–25; Arizona Department of Public Safety/AP Images, pp. 42–43; Astrid Galvan/AP Images, pp. 10–11; BLACKWHITEPAILYN/Shutterstock Images, pp. 12–13; Bruce Davis/Flickr, pp. 38–39; brunocoelho/Shutterstock Images, p. 46; Charles T. Peden/Shutterstock Images, pp. 6–7; Clara_Sh/Shutterstock Images, pp. 36–37; EcoPrint/Shutterstock Images, p. 1 (scorpion); Evgeny Subbotsky/Shutterstock Images, p. 44; Gary C. Tognoni/Shutterstock Images, pp. 40–41; jrituc-ci/Shutterstock Images, p. 1 (desert canyon); Jim_Brown_Photography/Shutterstock Images, pp. 32–33; Johnny Coate/Shutterstock Images, pp. 4–5; Jonathan Manjeot/Shutterstock Images, pp. 8–9; Kaisove/Wikimedia Commons, pp. 20–21; Lapa Smile/Shutterstock Images, pp. 30–31; Revilo Lessen/Shutterstock Images, pp. 28–29; Staff Sgt. Marcus Fichtl/Wikimedia Commons, pp. 16–17; stock.film/Shutterstock Images, pp. 34–35; Vaclav Mach/Shutterstock Images, pp. 22–23; Wirestock Creators/Adobe Stock, pp. 26–27
Design Elements: Nik Merkulov/Shutterstock Images (grunge background); queezz/Shutterstock Images (brush texture)

Library of Congress Control Number: 2023939327

Publisher's Cataloging-in-Publication Data
Names: Rusick, Jessica, author.
Title: Surviving the desert / by Jessica Rusick
Description: Minneapolis, Minnesota : Abdo Publishing, 2024 | Series: Xtreme survival | Includes online resources and index.
Identifiers: ISBN 9781098291822 (lib. bdg.) | ISBN 9781098278724 (ebook)
Subjects: LCSH: Survival--Juvenile literature. | Survival skills--Juvenile literature. | Hot weather conditions--Juvenile literature. | Desert survival--Juvenile literature. | Outdoor life--Juvenile literature. | Wilderness survival--Juvenile literature.
Classification: DDC 613.69--dc23

TABLE OF CONTENTS

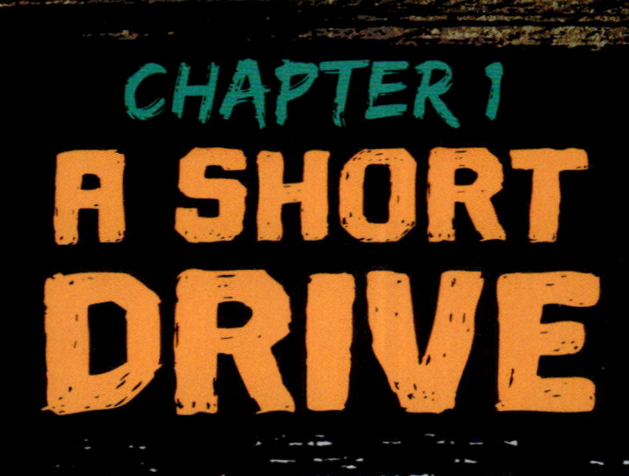

CHAPTER 1
A SHORT
DRIVE

On March 31, 2016, Ann Rodgers was driving to Phoenix, Arizona, to visit her grandchildren. The drive was supposed to take less than two hours, but Rodgers's car was low on fuel. She tried to find a gas station but became lost on a **secluded** desert road. There was no cell phone service to call for help. Soon her car ran out of fuel.

Rodgers was driving up Interstate 10 (*pictured*) from Tucson, Arizona, to Phoenix when her car died.

Rodgers stayed in her car overnight, where she had some food and water. She hoped someone would find her, but no one came. By the next morning, Rodgers was out of supplies. She decided to find help or a cell phone signal. Rodgers was **stranded** in the Arizona desert. How would she survive?

XTREME FACT

Desert survival experts say it's best to stay with your car. That's because a car is easier for rescuers to spot.

When Rodgers's car was found on April 3, helicopters and ground rescue started searching for her. Had she stayed near her car, rescuers would have found her much sooner.

Deserts receive less than
10 inches (25 cm) of rain per
year. Some areas get less than
0.1 inches (0.25 cm) of rain annually.

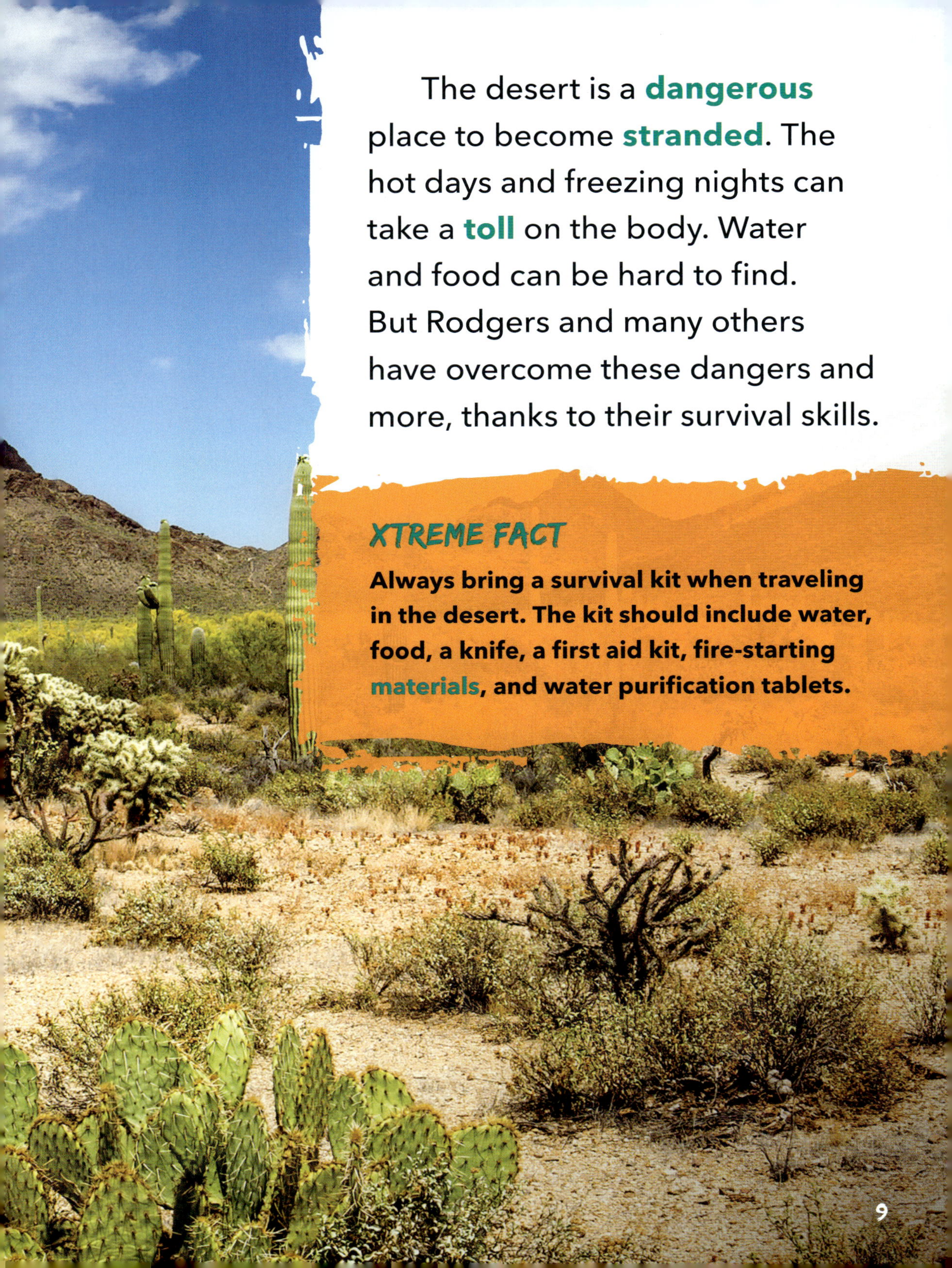

The desert is a **dangerous** place to become **stranded**. The hot days and freezing nights can take a **toll** on the body. Water and food can be hard to find. But Rodgers and many others have overcome these dangers and more, thanks to their survival skills.

SURVIVOR SPOTLIGHT

Rodgers drank pond water and ate desert plants to survive. At night, she built a fire for warmth. She made a "HELP" sign using rocks, sticks, and animal bones. Rodgers also left a note by the sign saying where she planned to walk. Rescuers saw the sign and note, which helped them find Rodgers. After nine days in the desert, she was rescued!

When Rodgers was stranded, she used the skills she had learned from survival classes and books.

CHAPTER 2

SURVIVING THE SUN

Desert survival requires protecting yourself from the heat and sun. Temperatures can reach 100 degrees Fahrenheit (38°C) or higher. At high temperatures, the body struggles to keep itself cool. This can lead to heatstroke, which causes headaches, nausea, seizures, and eventually death.

Staying cool and protected against heatstroke is important. A person can die from heatstroke in just a few hours.

The desert is hottest during the afternoon. So it's important to stay cool. Seek shade and avoid expending energy. In 2010, Ed Rosenthal got lost hiking in the California desert. For several days, he lay in the shade of a small canyon, only moving to avoid the sun. This helped Rosenthal stay alive until he was rescued.

If possible, avoid
lying directly on the
sand when the sun is
out. The sand can be
hotter than the air!

If you can't find shade, make it yourself. In 2018, hiker Claire Nelson fell and broke her pelvis in the California desert. She could barely move due to the pain. During the day, she was directly in the hot sun. Nelson formed a **makeshift** shelter using clothing and a hiking stick to cover her face and stay protected from heatstroke. Rescuers found her after four days!

If you don't have supplies
to create a shelter, gather
branches or brush to
create a shade structure.

CHAPTER 3

FINDING WATER

Finding water is one of the most important parts of desert survival. In hot temperatures, the body quickly becomes dehydrated, losing water and **electrolytes**. Dehydration causes thirst, confusion, weakness, and eventually death.

In 1994, Italian athlete Mauro Prosperi became lost in
the Sahara Desert in Africa while running a marathon. He
squeezed water from plant roots to stay hydrated. Water
may also be found in the ground beneath broad-leaved
plants. These require more water than other desert plants,
so they often grow in places with more groundwater.

Participants in the Marathon des Sables, known to be one of the toughest races on Earth. Prosperi became lost in the desert while running in this race.

Use a stick to dig in dry creek beds. Once you find water, let it flow into the hole and sit for a few minutes before you purify it to drink.

There are other ways to find water. Dew builds on rocks and plants in the morning. Use clothes to wipe up and wring the dew into your mouth. **Stagnant** water can be found by digging at the edges of dry creek beds. If possible, boil water or clean it with a purification tablet before drinking it.

Preventing dehydration also means preserving the water in your body. Humans lose water through sweat. Search for water in the mornings and evenings, when temperatures are cooler. You can also watch for birds and insects, as they usually live near water.

Following animal tracks can
lead you to a water source.

CHAPTER 4

FINDING FOOD

In the desert, finding water is more important than finding food. Humans can survive for weeks without food, so hunting is often not worth the energy. Instead, reserve energy for finding water and staying alive until rescuers find you. Do not look for food unless you have **ample** water.

Digestion requires water.
So you should not eat
unless you also have
something to drink.

There are several food sources in the desert. Prosperi caught lizards, snakes, and insects. Their blood also kept him hydrated. If possible, cook lizards and snakes before eating them. They carry bacteria that can make humans sick. Insects are safe to eat raw.

Insects such as locusts (*pictured*), honeypot ants, and beetles are good to eat. They contain a lot of protein and fat.

Although cactus water is toxic
to humans, several cactus fruits,
like the prickly pear, are safe
to eat. These fruits also contain
some drinkable water.

Many animals are
safe to eat, but avoid
those with bright colors.
They may be poisonous.
And don't search for
food under rocks or in
holes. You could get
bitten or stung!

CHAPTER 5
STAYING
WARM

At night, desert temperatures can drop below freezing. This can cause low body temperature, or hypothermia. People with hypothermia experience shivering, confusion, and **intense** sleepiness. Hypothermia can also cause death in hours.

Sheltering near rocks or under rock outcrops at night can help you keep warm since rocks retain heat from the day.

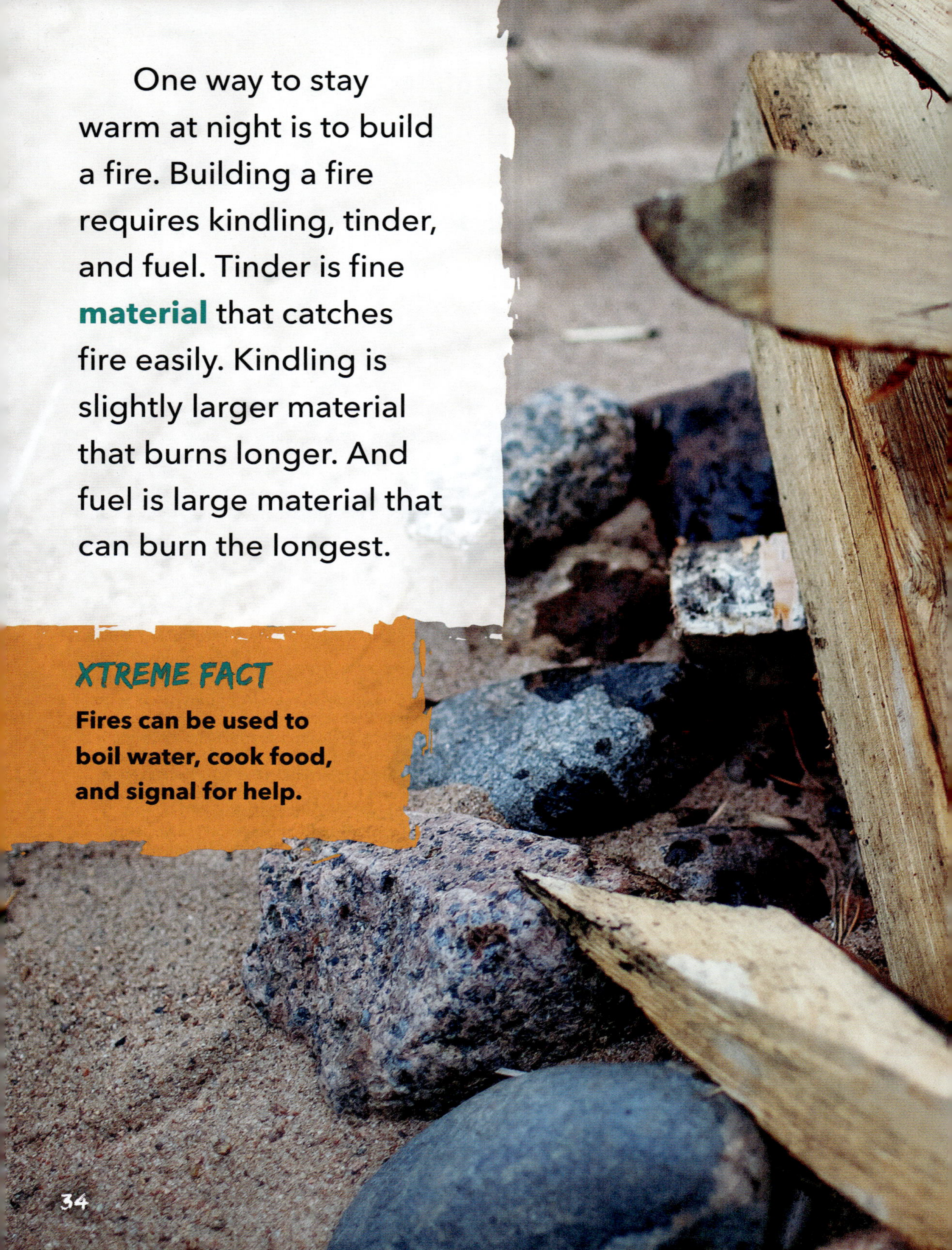

One way to stay warm at night is to build a fire. Building a fire requires kindling, tinder, and fuel. Tinder is fine **material** that catches fire easily. Kindling is slightly larger material that burns longer. And fuel is large material that can burn the longest.

Find or clear a spot where a fire will not spread and damage the surrounding area. If possible, create a fire ring with dirt and rocks to contain the fire.

Make sure you don't use all your fuel at once. Set some aside so you can keep your fire going through the night.

Desert tinder includes dry brush and animal droppings. Dead wood serves as kindling and fuel. Arrange the fuel in a tepee shape. Place the tinder and kindling underneath the fuel. Then light the tinder with a match or lighter.

XTREME FACT
You can use
eyeglasses or a
camera lens to
direct sunlight
onto tinder to
start a fire.

It's not always possible to build a fire. In 2012, hiker Victoria Grover broke her leg and was **stranded** in the Utah desert. She wrapped a poncho around her body and head at night. This captured the warmth of her breath and helped her avoid hypothermia. Grover also sat up instead of lying on the cold ground. She **focused** on staying awake. After multiple nights, Grover was rescued.

Grover survived in the Box Death Hollow Wilderness by sleeping during the day and staying awake at night to avoid hypothermia.

CHAPTER 6

GETTING RESCUED

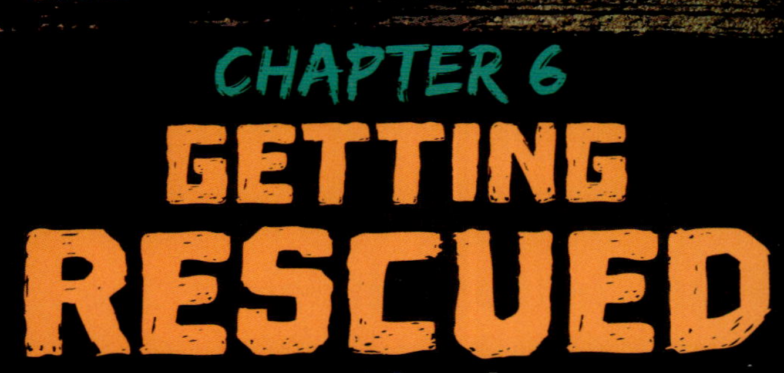

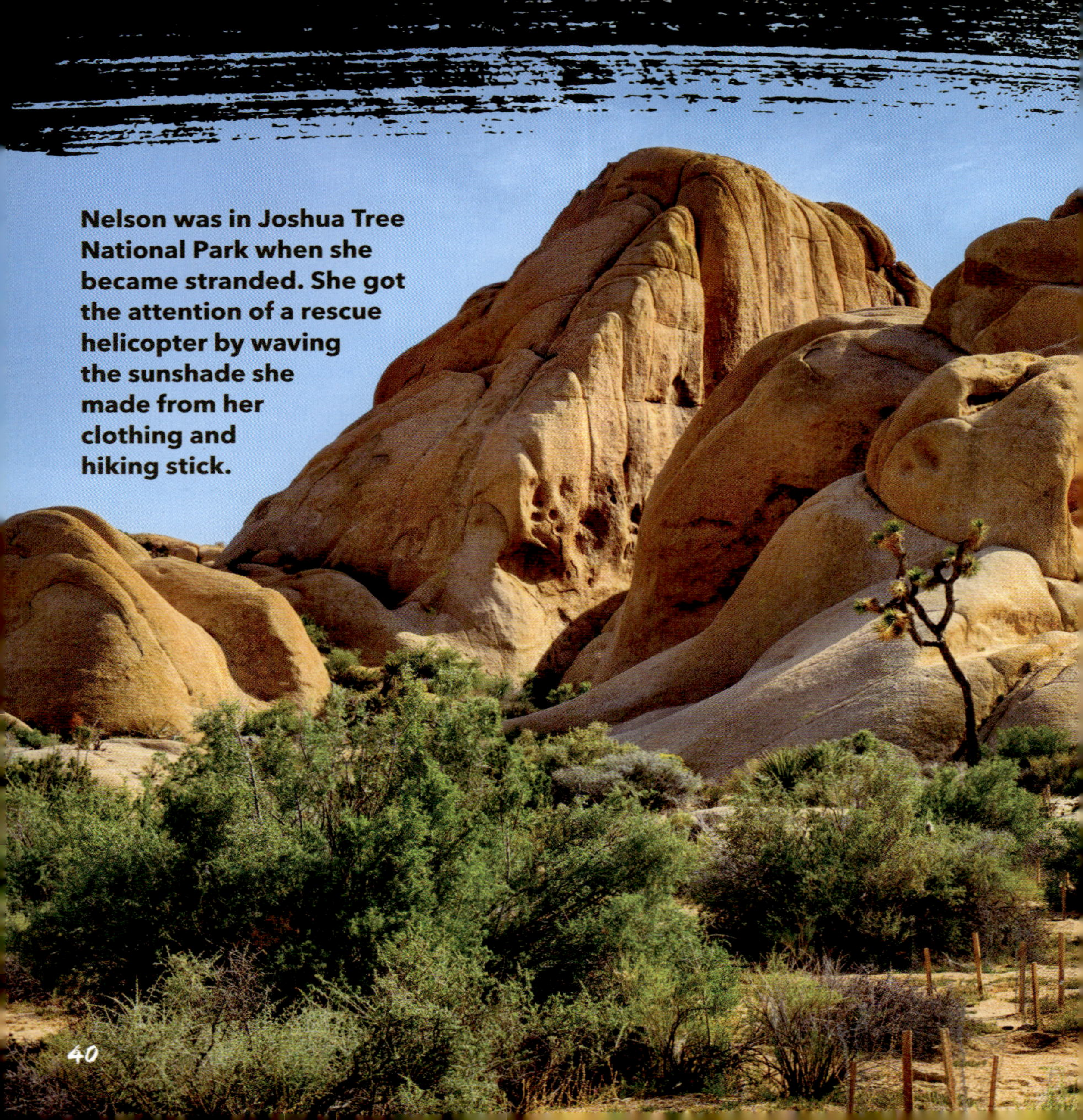

Nelson was in Joshua Tree National Park when she became stranded. She got the attention of a rescue helicopter by waving the sunshade she made from her clothing and hiking stick.

Being **stranded** in the desert is scary. It's important to remain calm and **focus** on getting rescued. Nelson was in **extreme** pain after breaking her pelvis. But she remained **optimistic**, focusing on protecting her body from the sun and seeing her family again.

XTREME FACT

Hiking **experts** say it's important to tell at least two people where you plan to go and when you expect to be back. That way, they can notify rescuers if you are missing.

Desert rescuers often search for missing people by helicopter. There are several ways to get noticed. Write "HELP" or "SOS" in large letters with sticks, rocks, and other **materials**. Signal by waving reflective objects or bright clothing. If possible, climb to higher ground. You can also seek help by looking for trails, dirt roads, or other signs of life that may lead to other people.

CHAPTER 7

SURVIVING THE DESERT

Being **stranded** in the desert means having little **access** to water, food, and shelter. **Extreme** temperatures are constant dangers. But with top-notch survival skills and a bit of luck, some people have overcome these challenges to survive. Think back on the desert survival stories you read and skills you learned. Could you survive in the desert?

When visiting the desert, wear a wide-brimmed hat, long sleeves, and pants. These will help protect you from the sun and cold, especially if you get stranded.

XTREME CHALLENGE

1) What are the three types of materials needed to build a fire?

2) What animals did Mauro Prosperi eat to survive in the desert?

3) If you were stranded in the desert, would you rather have a tarp to provide shade or matches to start a fire? Why?

4) What causes hypothermia?

5) How would you keep calm if you were stranded in the desert?

GLOSSARY

access—the opportunity to gain or use something.

ample—enough to satisfy a need.

dangerous—able or likely to cause hurt or harm.

electrolyte—an ion that regulates or affects metabolic processes of the body.

expert—a person very knowledgeable about a certain subject.

extreme—very much, or to a very great degree.

focus—to concentrate on or pay particular attention to.

intense—existing in an extreme degree.

makeshift—temporary or substitute.

material—what a thing is made up of.

optimistic—feeling or showing hope for the future.

secluded—cut off from others.

stagnant—still or not flowing.

stranded—lacking the means to leave a place.

toll—a cost in life or health.

ONLINE RESOURCES

To learn more about desert survival, please visit **abdobooklinks.com** or scan this QR code. These links are routinely monitored and updated to provide the most current information available.

INDEX